THE IDEAL AIM OF PHYSICAL SCIENCE

W0114630

CAMBRIDGE
UNIVERSITY PRESS

University Printing House, Cambridge CB2 8BS, United Kingdom

Cambridge University Press is part of the University of Cambridge.

It furthers the University's mission by disseminating knowledge in the pursuit of
education, learning and research at the highest international levels of excellence.

www.cambridge.org
Information on this title: www.cambridge.org/9781316619841

© Cambridge University Press 1925

This publication is in copyright. Subject to statutory exception
and to the provisions of relevant collective licensing agreements,
no reproduction of any part may take place without the written
permission of Cambridge University Press.

First published 1925
First paperback edition 2016

A catalogue record for this publication is available from the British Library

ISBN 978-1-316-61984-1 Paperback

Cambridge University Press has no responsibility for the persistence or accuracy
of URLs for external or third-party internet websites referred to in this publication,
and does not guarantee that any content on such websites is, or will remain,
accurate or appropriate.

THE IDEAL AIM OF PHYSICAL SCIENCE

*A Lecture delivered on November 7, 1924
before the University of London,
at King's College*

BY

E. W. HOBSON, Sc.D., LL.D., F.R.S.
Sadleirian Professor of Pure Mathematics

CAMBRIDGE
AT THE UNIVERSITY PRESS
1925

THE IDEAL AIM OF PHYSICAL SCIENCE

A SCRUTINY of the activities and of the utterances of the men who, through the centuries, have been the pioneers in scientific discovery does not lead to the conclusion that their aims were wholly, or even mainly, of what may be called the practical order. As we all know, their labours have resulted in a progressive transformation of industrial and social life in the modern age. But the true man of Science has been dominated by a deep-seated impulse to obtain a precise knowledge of, and in some sense to comprehend, natural phenomena. In fact his aim has been to obtain the satisfaction of what I may perhaps venture to describe as glorified curiosity. For him, the practical inventions to which I have alluded have been at most a by-product of his activities, subordinate to his striving after the attainment of a purely mental satisfaction. In fact, with some notable exceptions, men of Science have, in a large measure, relegated the exploitation for practical purposes of their discoveries to men of a different type of mind; to the inventors, whose labours were however absolutely dependent, for the possibility of success, upon the results of pure Science.

It is of the ideal aim of Physical Science, as distinct from the practical aim of controlling natural phenomena, and in particular of the different manners in which this ideal aim has been, and may be formulated, that I shall speak in this Lecture. I propose to illustrate the matter by means of references to some of the most modern physical theories. As a simplification, necessary in a single Lecture, I shall avoid all reference to those phenomena in which the fact that some of the objects in the physical world are what we call living organisms is, or appears to be, relevant. I shall, in fact, restrict myself to that simpler part of Natural Science which we call Physical Science, but which includes Chemistry, which is essentially a department of Physics, and also Mathematics, which has its origin in the same source as Physics in general, and is, in its abstract form, the indispensable instrument for the formulation of all physical theories which have reached an advanced stage in their development. Physics and Chemistry, owing to methodological and historical reasons, were for a long period cultivated separately, but they have in our time, owing to the discoveries in the last few decades, in the domain of Radio-activity, been forced to recognize, at least in principle, their essential unity.

The form in which the ideal aim of Physical Science has been conceived has varied greatly at different epochs, and it has depended very largely upon the general philosophical outlook of particular periods and upon that of different groups of thinkers during one and the same period. In recent times, in which specialization of knowledge has been much more prominent than in earlier periods, many, perhaps most, men of Science have been but little subject to the direct influence of the movements of speculative Thought; and they have often conceived the aims of Science in accordance with the uncritical views of what is called common sense. But the recent theory of relativity, promulgated by Einstein and his followers, subversive as it is of certain assumptions hitherto universally made both by Science and common sense, has forced upon the scientific world a reconsideration of fundamental matters, in such wise that a renewed discussion of the more precise formulation of the ideal aim of Physical Science can scarcely be avoided.

In making attempts to define the general aim of Physical Science as something which is, at least in some degree, capable of attainment, we are not bound to accept as satisfactory the notions of particular great men of Science, present or past, as to the fundamental character of such ideal aim. A man of Science may think he is engaged in doing one thing when in fact he is doing something that turns out, when closely scrutinized, to be quite different from what he imagined.

There is one source of frequent misunderstanding as to the aim of Science; that is an insufficient delineation of the frontier line between Physical Science and general Philosophy. Without such delineation the aims of Science may be so extended as to include some at least of the wider aims of Metaphysical Philosophy, the attainment of which requires something beyond those methods of procedure which are alone essential for the purposes of Physical Science.

Philosophy, in its eclectic review of the whole of human experience, has to take into account the methods and general results of scientific enquiry, and has to use them for the purpose of making ontological and other inferences applicable to the universal domain with which Philosophy is concerned.

The making of such inferences from the results of Science, and a scrutiny of the validity of such inferences, is no part of the direct aims of Physical Science, but the mistake is frequently made of including in those aims what should really be regarded as belonging to the aims of Philosophical thought, so far as it uses the results of Science as material for its own purposes. No doubt, like all frontiers, the frontier between Physical Science and Philosophy is partly conventional, but the position of Physical Science is greatly strengthened if its aim is so formulated that it is shielded from those elements of doubt and uncertainty which may arise in connection with the wider philosophical inferences from the results of Science. The aim of Physical Science should accordingly be regarded as a modest one, compared with that of general Philosophy. There is no need to make Science responsible for the inferences which Philosophy may make from scientific results, and least of all for the underlying ontological, or other, assumptions upon which such inferences may be based. It is said that Science has to deal with physical phenomena, but that every phenomenon is an appearance of something to someone; and thus that the characteristics of phenomena may be used for making inferences as to the characteristics of that something which gives rise to the appearances. Thus a real ground, or an underlying substance, or things in themselves, or νοούμενα are brought into relation with scientific results. Inferences of this species may be quite legitimate matters for general Philosophy, but there is, I think, an overwhelming advantage, for clarity of thinking, both on the philosophical and on the scientific side, if such inferences are completely extruded from the province of Physical Science, which may, with great advantage, confine its attention to dealing with the phenomena themselves, and can thus free itself from speculative accretions, the due discussion of which, important as it no doubt is in general Thought, requires methods and ideas which are not necessary for the primary purpose of classifying and systematizing the march of physical phenomena.

The notion that the main aim of Physical Science is to *explain* physical phenomena has, at least until recently, been a dominant conception of the function of Science. It is however necessary to enquire what precise meanings have been attached to the term "explanation." To explain, or make plain, a phenomenon, or a group of phenomena, has often been taken to denote the reduction of the relatively unfamiliar to a case of the relatively familiar. The making of mechanical models which imitate the elastic, or other, properties of matter is an example of the idea that phenomena can be explained by reducing them to an instance of what is so familiar that it is regarded as already plain; the assumption being made that the functioning of the model does not, at least provisionally, itself require elucidation.

Since the time of the Greeks the idea has been

extraordinarily persistent in the history of Physical Science that all physical interaction must be due to what takes place on the contact of bodies, or to their impingement on one another. The fact that, in order to produce physical changes in our environment, we place some portion of our bodies in contact with external things has made the notion of contact action so familiar to us that the reduction of physical phenomena of any class to a case of contact action affords, in the sense to which I am referring, an explanation of the phenomena in question. To this order of ideas the introduction into Physical Science of the various ethers and effluvia which have figured in the history of Physics is due; they were all introduced with a view to the mediation of contact action. When Newton's law of Gravitation was formulated, it was held by Newton and his contemporaries that, as action at a distance is inconceivable, the law itself afforded no explanation of gravitational phenomena; that, in fact, it was necessary, for a real explanation, to show that the formal law could be accounted for by action propagated through some medium by which the apparent, but unintelligible, action at a distance could be reduced to an intelligible contact action, or in accordance with Le Sage's theory, to a system of impacts.

The numerous unsuccessful attempts which were made to effect this reduction were inspired by this supposed need. The full recognition of the fact that action by contact or shock is in need of further elucidation; and that in fact a reduction to spatial scale of phenomena to be explained is all that can be accomplished by the introduction of an ether, or intervening medium, is even at the present time, in some quarters at least, incomplete. It is only right to emphasize the fact that the employment of substantial ethers has been of immense value at various stages in the development of Physical Science, although it only leads to an explanation of phenomena in the very circumscribed sense of the term "explanation" of which I am now speaking; moreover there are in general insuperable difficulties connected with the constitution and physical properties of the ethers or media.

The most rigorous sense in which the term explanation is used is that in which the major and minor premisses of a syllogism explain the conclusion; that is of reduction to logical identity. The application of this conception of what is meant by "explanation" to the case of physical phenomena has taken a very different form in the modern period from that which was prevalent in medieval times; but traces of the latter conception have still appeared from time to time up to our own day. We come thus to the notion of explanation by means of *à priori* laws. The idea that the laws of physical phenomena can be obtained as *à priori* constructions of pure thought; that physical phenomena must of necessity proceed in accordance with certain principles regarded as *à priori* necessary, dominated thinkers for a long time. Implications of this kind have been persistent even since the doctrine in its pure form has no longer been accepted. The doctrine in its fuller forms received its death blow after the Renaissance, when Bacon and others succeeded in establishing the view that empirical knowledge of physical facts and processes must be the ultimate and indispensable basis of the coordinative and constructive efforts of the human mind which lead to the edifice of scientific knowledge. That facts are the basis of all true Science has now been universally recognized, and has sometimes even been stated in such a manner as to depreciate unduly the importance of the other factors in the growth of Science, and to call for the warning expressed by Poincaré in the words "Science is built up of facts, as a house is built of stones; but an accumulation of facts is no more a Science than a heap of stones is a house." But the more ancient error of admitting scientific principles which have not been shown to rest upon the basis of actual physical experience dies hard. The principles of conservation of matter, of mass, and of energy, and the principle of least action, have all been based by various thinkers in modern times upon supposed necessities of thought. The notion of indestructibility, or conservation, as an *à priori* principle, has proved singularly unfortunate in relation to attempts to decide what it exactly is that is conserved during physical transformations. The principle of conservation of mass, of great value as a formulation of what holds good at least with sufficient approximation in a large class of physical processes, is now, in accordance with recently acquired knowledge, regarded as untenable in the absolute sense in which it had been accepted. Indeed, in accordance with the theory of relativity, mass has been merged in energy, and both mass and energy have measures only of a relative kind. By Leibniz, the principle of the conservation of kinetic energy was made to rest upon the *à priori* principle of causation; and by J. R. Mayer, and by Colding, both pioneers of the general principle of conservation of energy, the truth of the principle was not made to rest upon experimental evidence such as that adduced by Joule, but rather upon *à priori* principles relating to the necessary indestructibility of the original endowments of matter.

Even Clerk Maxwell, whose fame as a true scientific discoverer will be at least as enduring as that of any man of Science of the nineteenth century, made an attempt to prove the principle of inertia on an *à priori* ground.

But the main form which the notion that the function of Physical Science is to obtain explanations of physical phenomena and processes has taken in recent times is embodied in the idea that Physical Science must proceed, by stripping off from our sense data what is regarded as accident or subjective appearance, and penetrating to an underlying objective reality to which the phenomena or appearances are regarded as due. Thus the aim of Physical Science, from this point of view, is to obtain a knowledge of objective reality, or at least of that part of reality of which the physical world is the manifestation. The complete acceptance of empirical observation as the basis of Physical

Science would appear to have strengthened the opinion of the majority of modern men of Science that they are engaged in an attempt, which has been in some considerable measure successful, to obtain a knowledge of the characteristics of ultimate reality.

By what tests we are to be assured, at any particular stage of the evolution of Physical Science, or of any department of it, that we have at any point actually reached bed-rock reality is not clear. If the test is taken to consist of the reduction of physical sequences to a chain of causes and effects in which the relation of cause with effect becomes completely transparent, and appears as a relation not only of observed invariable sequence in point of fact, but as one in which any kind of necessity is to be discerned, it may, I believe, be safely asserted that in no single instance has this test been satisfied. The words of David Hume in criticism of the view that we are able to discover either efficient causation or logical necessity in the sequences of natural phenomena themselves have a weight of evidence in their favour which has, if possible, been rendered more cogent, and of application in a deepened sense, by the wonderful developments of Physical Science which have taken place since he wrote the following words: "When we look about us towards external objects, and consider the operations of causes, we are never able, in a single instance, to discover any power or necessary connection; any quality, which binds the effect to the cause, and renders the one an infallible consequence of the other. We can only find, that the one does actually in fact, follow the other."

The purely naïve realism which regards the objects of the physical world as having a reality completely consonant with, or identical with, what our sense-presentations directly exhibit is so easily refuted that it has rarely been held by men of Science or by philosophical thinkers, at least in modern times. The reality which is recognized by many men of Science of to-day involves a complete transformation into an unrecognizable form, of the real world of the naïve realist. To those men of Science of to-day who believe that Science leads the way to reality, the real world consists of such things as electrons, atomic nuclei, vibrations of atoms and molecules, radiation, and space-time. What the real world of such men of Science will be like a century hence it would be extremely interesting to know. It is assumed by those men of Science who hold this view that ultimate reality must be articulated in a manner precisely corresponding with the distinctions and postulations which modern Physical Science has for its own purposes constructed, or as they would prefer to say, has discerned in the physical domain.

Another view of the ideal aim of Physical Science has in recent times come into prominence, which is markedly divergent from views of the class which I have so far sketched, which may for purposes of reference be lumped together as the "explanatory" theory. The newer view to which I refer is frequently known as the "descriptive" theory of the nature of Physical Science; but this terminology is inadequate as a characterization of the view, and is indeed misleading, unless it is supplemented by an account, sufficiently precise, of the denotation of the term "descriptive"; I shall however for convenience use the term, in contradistinction from the "explanatory" view of which I have already spoken. The origin of the "descriptive" school of scientific thought is recent, or at least its explicit formulation. Amongst men of Science, Kirchhoff and Mach, and in this country K. Pearson, have been its chief advocates; but the germ of this view is to be found in the Positive Philosophy of Auguste Comte, and even in the critical Philosophy of Immanuel Kant. The denial by Kant of the possibility of any theoretical knowledge of things in themselves, or $\nu oo\acute{\nu}\mu\epsilon\nu a$, and thus the conception that all theoretical knowledge has reference to appearances only, is of significance in this connection.

In accordance with the descriptive theory it is not the function of Physical Science to obtain explanations of physical phenomena and their sequences, if the term explanation is used in any sense such as I have already described; and it is certainly not the function of Physical Science to refer phenomena to a real ground. The notion of causation as due to active agents, that is efficient causation, and the discovery within the phenomenal domain of rational necessity, are from the point of view of the descriptive view unnecessary to Physical Science, and are consequently extruded from it. The classification, reduction to laws, and finally the

subsumption under conceptual schemes, of perceptual physical sequences, is in accordance with
this view, the true function of Physical Science.
The term causation, when it is employed at all, is
taken merely to indicate the totality of conditions
which are found by observation to be in fact in-
variably associated with physical changes. The
term explanation, if it be used at all by an adherent
of this school, can only be taken to be synonymous
with systematization, that is, subsumption of the
phenomena to be explained, under a general law
or under the wider groups of laws which are known
as conceptual schemes or general scientific theories.
When a particular phenomenon or physical sequence has been linked up with other phenomena
or sequences, so that they are represented by means
of correlation with a purely conceptual scheme, it
has received the only kind of explanation which
Physical Science is capable of affording. According
to the descriptive theory as now held, and the
general correctness of which I maintain, all scientific
knowledge is absolutely dependent upon actual
physical experience, but the cooperation is required
of the human mind, working with sense-presentations as data, in the process of sifting and classifying the objects of sense and the sequences in
physical changes, and further, in the formulation
of rules or laws, in accordance with the results
of observation. In the later and higher stages of
the development of Science, abstract conceptual
schemes are constructed by means of the constructive imagination operating upon the earlier

generalizations of observation, for the purpose of
representation of phenomenal processes of some
more or less wide class. These schemes employ
conceptual objects which are subject to postulated
properties and relations with one another. The
perceptual objects of the physical world with which
Science has to deal have a certain degree of in-
dependence of particular percipients; otherwise
Science would not be public or intrasubjective, the
same for all percipients, and in this sense they have
objectivity. But whether, or in what sense, they
have an existence independent of all percipients,
or of any percipient, is a metaphysical question,
the answer to which is irrelevant to Physical Science,
although the contrary view is still stoutly maintained by many men of Science. No question as to
the reality, in any metaphysical sense, of objects or
processes presented by sensuous perception, is regarded, in accordance with the descriptive theory,
as relevant for the purposes of Physical Science.

In no single instance can it be effectively maintained that the ideal of the " explanatory " view has
been actually reached; that any phenomenon has
been explained in the full meaning of the term;
it would appear indeed that a full explanation of
anything would involve a complete explanation of
everything, so complete is apparently the intercon-
nection of things. At any stage reached by Science,
what purports to be an explanation, in the rigorous
sense, of any particular class of phenomena involves postulations which themselves present new
problems. In the search for full explanation, for

reality, Science is involved in an indefinite regress, in which the explanation obtained at each stage is itself in need of explanation, and this without end. The same difficulties, only on a different spatial or other scale, present themselves at every stage of the regress, and these difficulties are logically just as urgent as those of the original problem. This fact can easily be illustrated by means of instances connected with certain recent developments of Physical Science. The "descriptive" view of the nature of Physical Science is less sanguine than the "explanatory" view in its expectations as regards what Science can attain, but the efficiency of scientific knowledge as a means of coordinating our trains of physical experience, of representing large classes of physical sequences, of foretelling the existence of phenomenal facts not hitherto observed, and also of obtaining a practical domination of Nature, is in no way impaired. It has the advantage over the opposing theories of the nature of Physical Science of freedom from the implications of any metaphysical view of reality, idealistic or realistic; no small advantage when we take into account the vast divergence of view on such matters which has existed amongst thinkers of the past, and which continues undiminished in intensity at the present time.

In the later stages of the development of Physical Science, purely abstract conceptual schemes, as I have already said, are set up for the purpose of representation of phenomena of some more or less wide class; and these schemes employ conceptual objects which are subject to postulated systems of relations. Such conceptual schemes are of various degrees of abstractness; in many of them the conceptual objects employed are merely universals, which typify classes of perceptual objects. But the highest and most general scientific theories make use of conceptual objects to which no actual or perceptual objects directly correspond, or are known at a particular time to correspond. Such purely conceptual objects are constructs of the mind, operating by a process of idealization, generalization, and analysis, in which the constructive imagination is operative upon the streams of sense data. Such a conceptual theory is always of a hypothetical character, and it is subject to two great tests of validity. In the first place it must be logically coherent, or rational, in the sense that the postulations of the scheme do not contradict one another, and that the deductions from the postulations which are necessary for the functioning of the scheme are capable of being carried out by a purely logical procedure. In the second place the scheme must satisfy the test of providing an adequate representation or description, by means of a process of correlation, of the actual march of events in some larger or smaller class of physical phenomena to which the theory is designed to apply; and above all it must have the power of accurate *prediction* of phenomenal processes not before observed, but which fall within the class of sequences to which the theory is applicable.

A theory which satisfies these tests is always

liable to be superseded by another more general theory which is applicable to the representation of a wider class of phenomena than was the former. This process of supersession is normal in the history of Physical Science. The unification of what were earlier different branches of Science into a single branch is an essential element in the evolution of Science. A striking example of this process of unification, characterized by the construction of a conceptual scheme embracing a field of phenomena which had formerly been represented by separate schemes applicable to separate parts, is to be found in the theory of Electromagnetism, as developed by Faraday, Clerk Maxwell and others, during the nineteenth century. Previously, Light, Electricity, and Magnetism had been phenomena devoid of known connection with one another, and for which separate conceptual schemes, of varying degrees of sufficiency, had been set up. All these were superseded by the more general theory which embraced all three classes of phenomena under a single conceptual scheme. In our time this unification has proceeded still further; the unit electrical charge having become a fundamental concept in the modern theory of the atom.

From the point of view here maintained, the method of Physical Science is essentially pragmatic. One of our greatest men of Science has been credited with the statement that: "A scientific theory is not a dogma but a policy"; although I have not been able to verify the reference. Simplicity is a desirable characteristic of a scientific theory, and when it is sacrificed in the interest of generality of representative power, the simpler theory may be usefully retained for those purposes for which it suffices. Thus, I take it, the simpler gravitational theory of Newton, sufficing as it does for all ordinary astronomical purposes, will still be retained in all cases in which Einstein's theory does not afford any correction of sensible magnitude. The traditional theories of the measurement of space and time will also be retained for the ordinary purposes of Science.

The postulations of any theory of physical phenomena usually appear to possess a certain arbitrariness, and they in general present new problems relating to their reduction to simpler postulations, or to such as may be better adapted to further the process of amalgamation of the theory with theories which have to do with other kinds of phenomena. The conceptual postulations relating to electrons and atomic nuclei in the modern theory of the constitution of atoms form no exception to this statement. These symbolic concepts certainly invite further analysis, especially when we find prominent investigators in this field speaking of electrons being subject to distortion under internal forces.

The most precise and concise form of language which we possess is that of Mathematics, and it is therefore not surprising that all physical theories which have attained to a sufficient degree of precision require that language for their formulation. This fact is at once an illustration and a confirmation of the general correctness of what is called the "descriptive" theory of the nature of Physical Science. For the abstract Arithmetic, the abstract Geometry, and the abstract Mechanics, with which Mathematics is concerned, do not employ sensuous objects or physical processes as given by intuition, as immediately presented, but work with idealized objects in which the greater part of what is given by intuition has been removed by abstraction, and frequently also with objects which, by a process of generalization, have been so constructed that, removed from the intuitional region, they do not even lend themselves to any approximate representation of anything directly accessible to the sensuous imagination. Their representative power is of an indirect character. Of this character is, for example, the quadruply ordered manifold which is fundamental in Einstein's theory of relativity, the "world" of Minkowski. This manifold can be conceived by the mind, but cannot be imagined as an object of sense. To ascribe to it any other kind of reality than that which appertains to every permanent concept as such, would appear to be at least otiose, and is certainly unnecessary for the attainment of the purposes for which the concept has been constructed.

I intend to devote the remainder of this Lecture to some observations, necessarily brief and summary, relating to the chief physical theories which hold the field at the present time, and which illustrate clearly the true character, as conceptual schemes, of physical theories in general. In this connection the evolution of the ancient concepts of space, time, and matter, and their unification in a new conceptual form, is made manifest. Considerations of time unfortunately compel me to omit even a cursory historical account of the growth of these concepts. Such an historical retrospect of the main features in this evolution is not only of the highest historical interest as a Chapter in the history of human thought, but it is of great value in the elucidation of the forms in which these concepts have a place in modern Science.

There have appeared recently many writings containing a more or less popular discussion of the new ideas on the measurement of time and space, and on kinetics, which are associated chiefly with the name of Einstein; and these frequently contain a characterization of the differences between Einstein's theory and the Classical Dynamics associated with the names of Newton and Galileo, and which was developed in a complete explicit form by Newton. Some of these popular presentations have, I think, tended to produce a certain amount of misunderstanding as regards the exact nature of the differences between the relativistic theory and the Classical Dynamics. It is, for example, sometimes stated, or implied, that the Newtonian system rests upon a basic conception of absolute space and time, whereas the Einstein system involves a thoroughgoing relativistic conception of both; that in fact all motion is absolute in the one system and relative in the other. Such a statement is, I think, partly erroneous, and partly incomplete; it is in any case quite insufficient. I shall accordingly discuss, as briefly as possible, what appear to be the basic differences between the two schemes, and also their similarities. In any characterization of the Newtonian scheme it is not desirable to be bound too closely by Newton's own presentation of it, but rather to present it in a form which takes account of the critical work on the subject subsequent to the time of Newton. As Newtonian Dynamics will no doubt continue to be employed for the many purposes for which it suffices, it is, apart from the comparison with Einstein's scheme, of importance to possess a clear statement of it as a conceptual scheme.

Much of the difficulty of understanding the precise character of such schemes arises from the habit, which many exponents have, of using language which leaves readers in the dark, as to whether the writer is speaking of concepts or of percepts, of abstract geometrical space or abstract time, or of actual or intuitional space or time, as given directly in experience. For clarity, the abstract or conceptual side of such matters must be clearly distinguished from that side which is concerned with the application of the theory to actual physical measurements. This is, of course, in accord with the descriptive view of the nature of Physical Science; and in the particular connection with which I am concerned, that view receives most cogent support from an analysis of the theories involved.

Each individual has his own private temporal

4

and spatial intuitions, derived by an incipient process of abstraction, or distinction, from the originally undifferentiated stream of his sensuous experience. Leaving the intuitional space and duration of the individual experient as analyzed by Psychologists, we pass on to the constructs of physical space, and of physical or public time, of which the origin is to be found in intrasubjective intercourse. It is in physical space and public time that actual spatial and temporal measurements are made, both in practical life and for the ordinary purposes of Science.

The measurements of physical space are made by means of measuring rods or other instruments, and also by the less direct processes in which use is made of optical phenomena.

The measurement of public time is based upon some standard spatially measurable process, often that of the rotation of the earth. There are many such processes, found by experience to be very approximately congruent. The ordinary units of time, the year, the mean solar day, the minute, and the second are obtained, as units of public time, by a somewhat complicated construction which it is unnecessary to describe here.

The essential property of physical time and space, considered as constructed schemes of measurement, is that they can be correlated with the private spaces and times of any particular individual. It had, before the rise of the relativity theory, always been assumed that these constructions of a single physical space, and a single public time, independent of each other, would suffice for all purposes, and for all observers under all circumstances. The assumption seemed so obviously a matter of course, so axiomatic, that it remained unnoticed that any assumption, requiring verification, had been made.

The first great breach of the Einstein theory with the previously universally accepted tradition consists in a denial of the sufficiency for all purposes of these constructs, a single physical space and a single public time, independent of each other, as affording the basis of a system of spatial and temporal measurements which will completely accord with the spatio-temporal experiences of all observers under all circumstances. In order to see how the absence of a single physical space and the single public time are compensated for in the Einstein theory we must proceed to consider the purely conceptual basis of the systems of actual spatio-temporal measurement, first that which has been universally applied before the rise of the Einstein scheme, and in particular lies at the base of the Classical Dynamics of Galileo and Newton; and secondly the system which is fundamental in the Einstein scheme.

In the traditional system, absolute, or geometrical space is a three-fold ordered manifold of ideal objects or elements (the points of abstract geometry) in which each object is correlated with three numbers, so that the relations of order of any element with other elements are uniquely specified by means of these numbers or coordinates. This order is purely abstract, neither specifically spatial, nor specifically temporal; our intuitional notions of spatial and of temporal order, from which the notion of order was derived, having been removed by abstraction. Before this manifold becomes a metric geometrical space, a metric system must be imposed upon it. This, as was shown by Riemann, can be done in an indefinitely great number of ways, but at present we are concerned with one only of these ways, that in which the separation, or distance, between any two elements is defined as the square root of the sum of the squares of the differences of the corresponding coordinates of the two elements. With this Pythagorean definition the whole system of Geometry with a Euclidean metric can be developed. This geometrical space contains an absolute frame of reference which may be regarded as constituted by the three sets of elements in each of which two of the defining numbers, or coordinates, have the value zero, the third coordinate having all values, positive or negative.

In a similar manner, abstract time is represented by a singly-ordered set of numbers, and this is also an absolute system, the origin of reference being the time represented by the number zero. Here, as in the case of space, the intuitional notion of direction has been transcended; for example, the qualitative difference between past and future has been removed; the sole element of our temporal intuition which remains is that of simple order.

When Newton suggested the apparently circular conception of absolute time, as that which flows uniformly, we can only take it as a definition of absolute time, in which an interval is defined to be the difference of two numbers of the simply ordered manifold, the arithmetic continuum.

In order to introduce into this abstract scheme the abstract conception of motion, we conceive a particle to be an ideal object represented by three variable coordinates each of which is a continuous function of the time variable. Moreover, we conceive to be associated with this particle a fixed number called its mass-number.

Time does not allow me to specify in detail the mode in which an ideal body may be conceived; it must suffice to say that it is an aggregation of ideal particles, whose separate motions are such as to satisfy certain conditions, which may be so chosen that any two particles of it have a distance from one another which is independent of the time-variable; and in this case the body is an ideal rigid body. A fixed mass-number also is associated with such an ideal body.

An ideal particle or body is defined to be at rest when its coordinates do not vary with the time-variable, and it is defined to be in motion when they do so vary; thus all rest or motion in this abstract scheme is by definition measured in relation to the absolute frame of reference, and may consequently be said to be absolute. In this absolute system of Dynamics the laws of motion, as formulated by Newton, must be taken to be in part definitions and in part postulations of the kind of motions to be imposed upon the ideal particles or bodies. The first two laws of motion amount to a definition of force, as measured by the product of the mass-number of an ideal particle into its acceleration (measured from the absolute frame). The third law, that of action and reaction, postulates certain relations between the mass-numbers of particles and their accelerations.

The efficiency of this conceptual scheme for the purpose of representing and predicting the motions of actual bodies in physical space depends upon the ascertained fact that it is possible to determine material frames of reference in physical space, so that, with a sufficient degree of approximation, they may be correlated with the absolute frame of the conceptual scheme, and thus that the motions of actual bodies, relatively to such a material frame, may be represented by the motions of ideal bodies in the conceptual system. A material frame which satisfies this condition determines what is known as a set of Newtonian axes, or an inertial frame. A material frame of this kind need only suffice for the particular purpose on hand; for some purposes a pair of straight lines on the ground and a plumb-line will be sufficient to provide the frame. In more delicate observations in which the solution of the earth is relevant, this will no longer be an adequate Newtonian frame. It is not possible to determine a material frame which for all possible purposes, and especially for all time, may be correlated with the absolute frame, without the introduction of "fictitious" forces.

The verification of the sufficiency of the scheme

of classical absolute Dynamics, for the determination of motions, as measured in physical space and with public time, can never be more than approximate, and it fails completely in certain cases. For verification, the Newtonian scheme must be taken as a whole; it is not possible to verify separately such a part of it as the principle of inertia. For such verification would involve observations under circumstances which are never realized. No actual body can ever be ascertained to be under no forces; when it is stated that an actual body is at rest, or in motion, uniform or other, in a straight line, or in rotation, the statement is meaningless unless a particular material frame of reference is assigned or implied. The principle of inertia is neither an *à priori* truth, nor a direct result of observation; it is a part of a hypothetical conceptual scheme, the utility of which, as a whole, must be judged by its success in fulfilling the functions for which it was devised. In any particular case the motions of bodies, as calculated from the scheme, depend upon the postulations of the scheme, and additional postulations, such, for example, as the law of gravitation, that are introduced for the particular purpose on hand. The calculated detailed motions of the conceptual bodies are the necessary logical consequences of the totality of these postulations, but this necessity cannot be ejected into any actual system in physical space; it is only actual experience which tells us whether or how far the motions of the actual bodies will be sufficiently described by those of the conceptual bodies.

Thus, neither logical necessity nor efficient causation is to be discerned in the sequences of motions of actual bodies.

It will be observed that, in the Dynamics of Galileo and Newton, space and time are treated as fixed frames into which physical phenomena are fitted, the theories of the measurement of space and time themselves being quite independent of the laws of Physics. Thus the Sciences of Geometry and Kinematics were regarded, even when they were recognized, which was far from being always the case, as in their origin dependent upon physical percepts, as independent of purely Physical Sciences, such as Optics, of which the laws presupposed in their statement the independent properties of spatial and temporal relations. The origin of the idea, now essential in the Einstein scheme, that the laws of Geometry and Kinematics are not independent of the laws of Physics, is to be found in the researches of Riemann and Helmholtz on the foundations of Geometry; and it is from the insight of these writers, and especially from the developments of Riemann, that the present views as to the solidarity of the theory of spatio-temporal measurements with Physics in general have ultimately arisen.

Before I pass on from the subject of Classical Dynamics I must refer to one last attempt to determine a frame of reference in physical space which might be regarded as absolute, in the sense of being independent of any ordinary material object. The notion that the electromagnetic phenomena of light take place in a substantial ether, through which material bodies can move freely without causing disturbance, led very naturally to the idea that this ether might itself provide the requisite frame with respect to which all motions of actual bodies might be measured, such a frame determining a permanent Newtonian system of axes. This attempt became abortive when the failure was established of the celebrated Michelson-Morley experiment to indicate the anticipated effect of the motion of the earth relatively to the ether. It was from this failure and from the discussions arising out of it that the Einstein scheme, involving a new Dynamics, came into being.

In the Einstein scheme, a single four-fold ordered manifold is the absolute construct which takes the place of the two absolute constructs of the Classical Dynamics, the three-fold ordered manifold of geometry and the singly-ordered manifold of the time-variable. Thus both the Einstein scheme and the Newtonian are based upon absolutes, in the one case the manifold of space-time, and in the other the manifold of absolute space together with the manifold of absolute time. In both systems there is relativity in all motions of actual bodies. The thoroughgoing relativity of the Einstein scheme makes its appearance in the diversity of the manner in which the fundamental four-fold is applied to represent actual spatial and temporal measurements made by an observer who uses a particular material body of reference and a material timekeeper. Thus space and time cannot be separated out from one another, in any way which is the same for all physical frames of reference. To enter into the details of the scheme is manifestly impossible for me, and indeed unnecessary, in view of the immense number of accounts of it which are readily accessible. I must confine myself to a reference to one or two points in which some of these accounts are such as to be liable to cause misunderstandings as to the fundamental nature of the scheme. In the classical scheme there is imposed upon the three-fold ordered manifold a particular metric system, the Euclidean, by which it becomes a Euclidean space, or more accurately geometrical space with a Euclidean metric. Einstein, on the other hand, imposes upon the fundamental four-fold a metric of a much more complicated character. The success of his scheme, as affording a conceptual representation of gravitational phenomena, depends upon the fact that he has shown how this metric can be so chosen that, when interpreted by an observer employing a material frame of reference, the motional phenomena in the field of material bodies which were formerly regarded as caused by gravitational forces, appear merely as the spatio-temporal measurements which correspond to the metric in the fundamental four-fold; they thus appear as phenomena dependent upon a metric system localized spatially and temporarily, not as caused by some supposed action at a distance. Matter, as revealed by its gravitational phenomena, manifests itself solely as a system of singularities in the metric system. There is no place left for the notion of

substance, either material or ethereal. It is sometimes said that space-time is crinkled or warped;
this statement is liable to produce misunderstanding, at least when it is not properly qualified, and is
in any case quite inappropriate. The fundamental
manifold has no properties intrinsic to itself, except
that of being ordered in accordance with an abstract
conception of order. Hence it has no singularities
which could be described as crinkles or warps. The
singularities by which matter is manifested belong
to the metric imposed upon the fundamental manifold from without, and are not intrinsic to the
manifold itself. Thus the statement to which I have
referred can only be taken to have a meaning when
the term space-time is taken to denote not the
manifold itself, but the imposed metric extrinsic to
the manifold, and chosen for certain purposes; and
such expressions as crinkles or warps, which refer
to sensuous impressions, are exceedingly inappropriate when taken to refer to such an abstract
conception as a metric.

To say that space-time, as so defined, as a kind
of conceptual scaffolding in an abstract scheme, is
the real world, the reality which is the ultimate
constitution of the world of experience, is a metaphysical assumption irrelevant to Physical Science,
and goes far beyond anything that has been, or can
be, established. This view of the matter is strengthened when we turn to certain other contemporary
lines of physical investigation which give strong
indications of limitations to which the representative power of the Einstein theory is subject, at

least in its present form, and even if the present attempts to extend it, so that it may cover the wider field of electromagnetics, are crowned with success.

The theory of Quanta has arisen in consequence of the apparent impossibility of representing various small scale phenomena, especially those of radiation, by means of the Classical Dynamics. It is essentially a scheme in which changes take place not continuously but by finite amounts, in which indivisible unit amounts of energy are involved. The Einstein Dynamics resembles the Classical Dynamics in being a continuous scheme, in which, at least in its original form, there is no room for the conception of Quanta. The theory of Quanta has been employed with a large degree of success in the theory of the Bohr-Rutherford constitution of the atom. The atom of a particular element, in accordance with this theory, consists in part of a minute nucleus, consisting of a certain number of hydrogen nuclei or of helium nuclei, together with a certain number of electrons, and in part of a planetary system of a certain number of electrons at distances from the nucleus very large compared with its size, moving in orbits round the nucleus, like the planets round the sun, under mutual forces which correspond to the Newtonian gravitational forces in the solar system. As long as these orbits remain stable they are calculated in accordance with the principles of continuous Dynamics. But from time to time instability sets in, and the system takes up by a sudden process, the exact nature of which is at present unanalyzed, a new stable system of orbits. How is the Einstein scheme to be reconciled with a scheme under which such phenomena as these can be conceived? We have, facing us to-day, the old antinomy of the continuous and the discrete; a continuous Dynamics, and an atomistic theory of energy. Even if the Einstein theory can be applied to give an account of the motions of the electrons in the atom during a period of stable motion, interpreting the electrical forces between the components of the atom in a manner analogous to that in which it interprets gravitational forces; how can it possibly be applied to represent the sudden changes from one type of continuous motion to another? This is the problem with which we are confronted, and which must be solved if the Einstein scheme is to take its place in a unified theory of physical phenomena. Must the continuous theory of space-time on which it is based be replaced by a theory in which the elements of space-time form a manifold which is no longer continuous, but is discrete?

Will the Science of the future involve a return to the conception of the representation of physical phenomena by a scheme in which thoroughgoing discreteness is an essential element, a discreteness more radical than that involved in the ancient atomism?

Are electrons and atomic nuclei more than parts of a temporary scaffolding, ultimately to disappear, of which there are so many examples in the history of Physical Science, and of which the substantial ether is a striking example? Will the whole theories

of chemical combinations and reactions, of electromagnetism, and of gravitation, be one day welded
together into a single conceptual scheme, of such a
character that these departments of Science will
have been absorbed into a single deductive Science
such as the special department of Geometry has
already become? These are all questions, which
we may hope that our successors will elucidate.

For EU product safety concerns, contact us at Calle de José Abascal, 56–1°, 28003 Madrid, Spain or eugpsr@cambridge.org.

12 301

www.ingramcontent.com/pod-product-compliance
Ingram Content Group UK Ltd.
Pitfield, Milton Keynes, MK11 3LW, UK
UKHW012312141225
465965UK00001B/3